HALLOWEEN ASTRONOMY

Mysteries of the Autumn Sky

JOHN A. READ

&

JENNIFER READ

STELLAR PUBLISHING

Halifax

For Derek Sweet, for giving me such a hard time about writing this.

Cover Design: Jennifer Read

Cover: Photo: Adobe Stock

Executive Editor: Kurtis Anstey

ISBN: 978-1-7327261-1-6

    1. Astronomy--Observers' manuals--Juvenile literature. 2. Astronomy--Amateurs' manuals--Juvenile literature. 3. Telescopes--Amateurs' manuals--Juvenile literature. I. Title. II. Title: Halloween Astronomy

PHOTO CREDITS

Telescope view source files for deep-sky objects were constructed from actual photos taken by the author, either using his personal four-inch refractor, twelve-inch Dobsonian and eight-inch Dobsonian, or using the following remote observatories: Abbey Ridge Observatory (owned by Dave Lane), and the Burke-Gaffney Observatory at Saint Mary's University, Halifax.

Other images used include: images from NASA which follow NASA's photo usage guidelines. Other image credits, including those for Hubble images can be found at the rear of the book.

Image of Celestron FirstScope Dobsonian compliments of Celestron; image of Explore Scientific FirstLight refractor compliments of Explore Scientific.

Star maps used in this book were sourced using Stellarium, an open-source stargazing program. These maps were then customized for the purpose of this book. Several of artist Johan Meuris constellation images from Stellarium are included in this book and usage rights can be found here: artlibre.org/licence/lal/en/.

TABLE OF CONTENTS

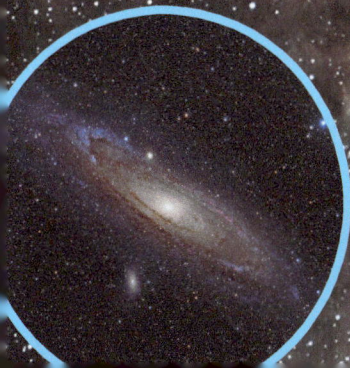

Halloween Skies

These are the skies as they appear on Halloween evening. How many constellations do you recognise?

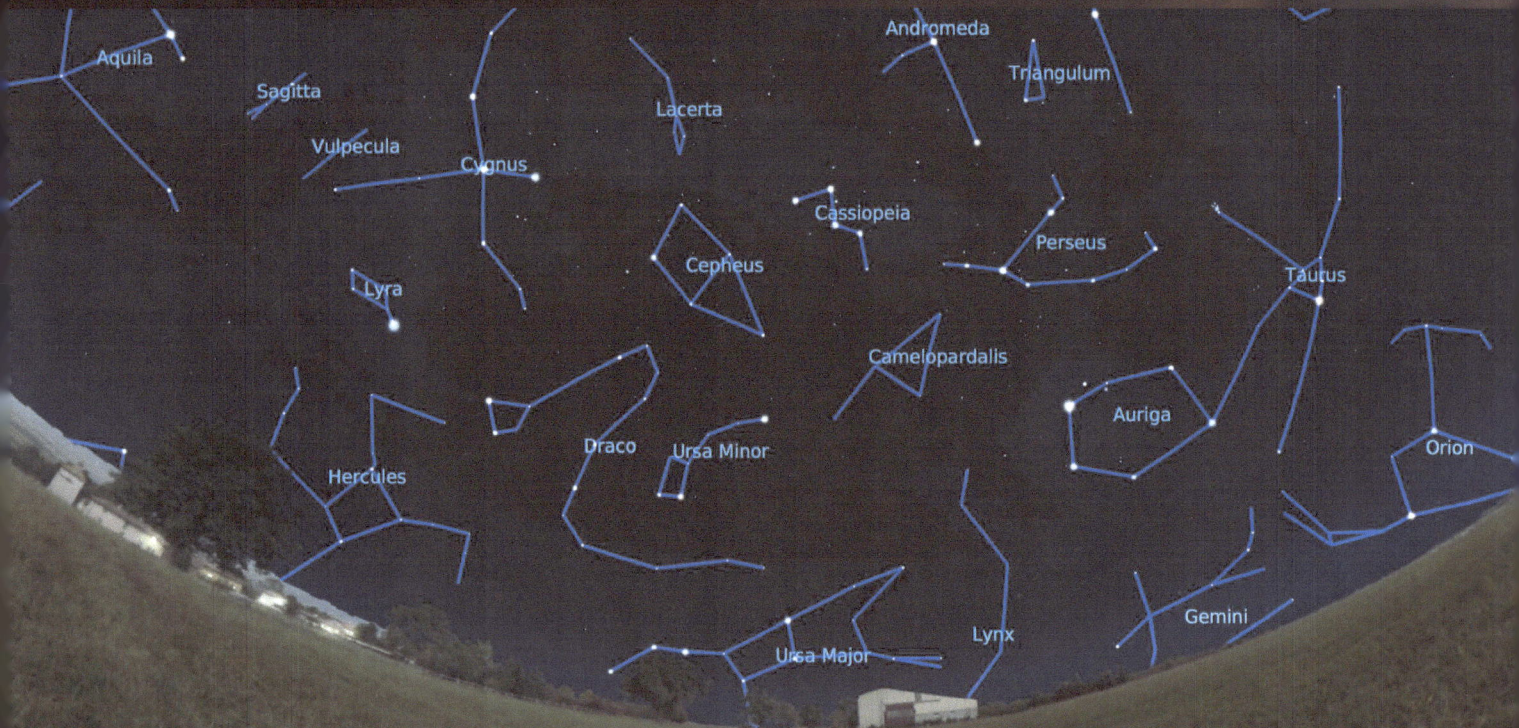

Aquila

Sagitta

Vulpecula

Cygnus

Lacerta

Andromeda

Triangulum

Lyra

Cepheus

Cassiopeia

Perseus

Taurus

Camelopardalis

Auriga

Orion

Draco

Ursa Minor

Hercules

Lynx

Gemini

Ursa Major

Looking North

Looking South

On the Moon, the terminator creeps westward and sunlight trickles between mountains and into valleys where ancient lava flows carved the landscape.

Far side of the Moon imaged by the Lunar Reconnaissance Orbiter (LRO)

Dark Side of the Moon?

Despite what the rock band Pink Floyd would have us believe, there is no "dark side" of the moon. The astronomical community refers to the side we can't see as the far side of the moon. This is the little over 40% that we can't see from Earth.

TELESCOPE VIEW

The Terminator

The **terminator** divides day and night on the Moon. At the terminator, the mountains cast long shadows and sunlight spills into craters, lighting one side, but not the other.

Lunokhod 1

5 months after Neil Armstrong and Buzz Aldrin landed on the Moon, the Soviet Union landed a rover. This rover, named Lunokhod 1, operated on the lunar surface for over 10 months!

Through a telescope, the crescent Venus looks deceptively like the Moon. Its featureless cloudy surface covers a sweltering and rugged landscape.

TELESCOPE VIEW

Venus was visited by NASA's Mariner 10 probe in 1974

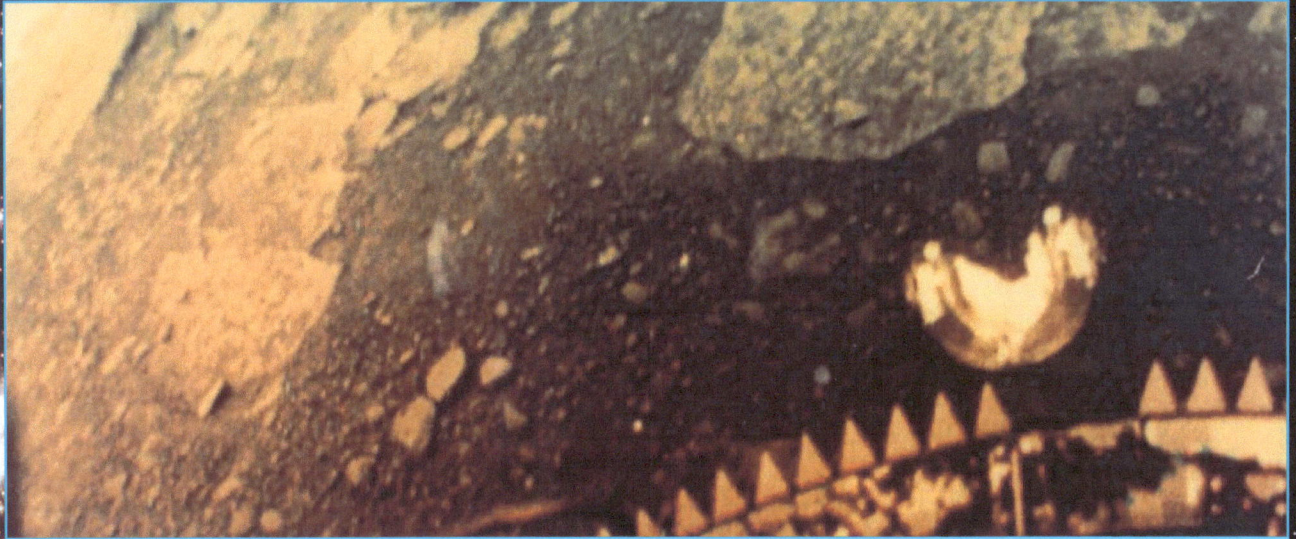

SURFACE OF VENUS IMAGED BY SOVIET SPACECRAFT VENERA 13

Fun Facts

Venus was named after the Roman goddess of love, but it would be hard for a planet to be more inhospitable. Venera 13 lasted only 127 minutes on its surface before succumbing to intense heat and pressure.

On Mars, a storm brews. Statically charged dust rises from the surface, collecting on the various robots that roam around this barren planet.

Mars Rover

TELESCOPE VIEW

Fun Fact

Martian storms can have extremely high wind speeds, up to 100 kph, but there is so little air pressure that the winds themselves can't harm any mechanical equipment on the surface.

How to Find Planets

As the planets in our Solar System orbit the Sun, they appear to wander across our night sky. For this reason, you'll need to use astronomy software to identify their current location. Free astronomy software is found at www.Stellarium.org

Valles Marineris
The largest canyon in the Solar System

Galilean moon Io slips behind the gas giant Jupiter, only to reappear a few hours later.

INNERMOST MOON IS USUALLY IO

Telescope/Binocular View

Jupiter can be seen in the evening sky for about half the year. Binoculars or a small telescope will reveal the planet as well as the four Galilean moons (Europa, Ganymede, Callisto, and Io).

Io

The Juno Spacecraft was named for Jupiter's wife

This red spot is a storm in Jupiter's atmosphere. It is twice as wide as the entire Earth!

WITH OVER 400 ACTIVE VOLCANOES, IO'S SURFACE IS CONSTANTLY CHANGING

Saturn's moon Titan is a world full of wonderous horrors. It rains methane. A single robot, named Huygens, lies frozen, yet not forgotten, on its surface.

Fun Facts

Titan is so cold that methane, which is usually a gas on Earth, is a liquid. The methane cycle on Titan is very similar to the water cycle on Earth. When the sun becomes a red giant and Earth is no longer habitable, Titan may become habitable for human life.

TELESCOPE VIEW

Titan

Cassini Spacecraft

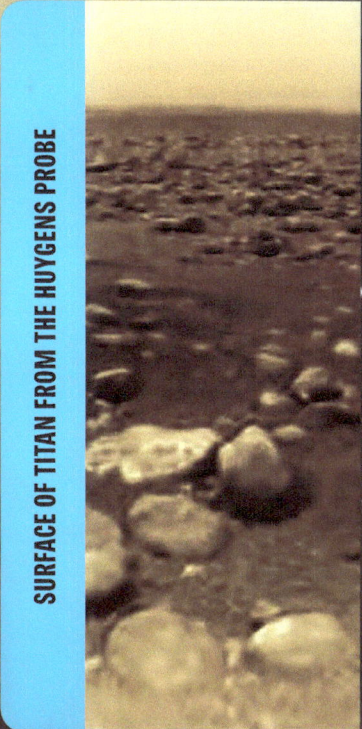

In Cassiopeia, a cluster sports a costume of its own. Sometimes a dragonfly with wings spread, or E.T. the extra-terrestrial with outspread arms. A fitting view for Halloween.

NGC 457 - Better known as the
Dragonfly Cluster

Cassiopeia

Andromeda's mother, infamous for her vanity, is often depicted
as holding a hand mirror. According to myth, she was placed
in the heavens bound the same way that her daughter was, as
punishment for the ordeal she put Andromeda through.

You can see it coming. A collision in the making. The Andromeda Galaxy approaches our home galaxy at almost 400,000 kilometers per hour. Fortunately, four billion years will pass before it arrives.

TELESCOPE VIEW

Perseus vs. Cetus

Andromeda was a princess of unparallelled beauty. In order to appease the jealous gods, she was chained to a rock at sea to be devoured by the sea monster Cetus. At the last moment, Perseus flew in on the Pegasus and used Medusa's head to turn Cetus to stone and rescue Andromeda.

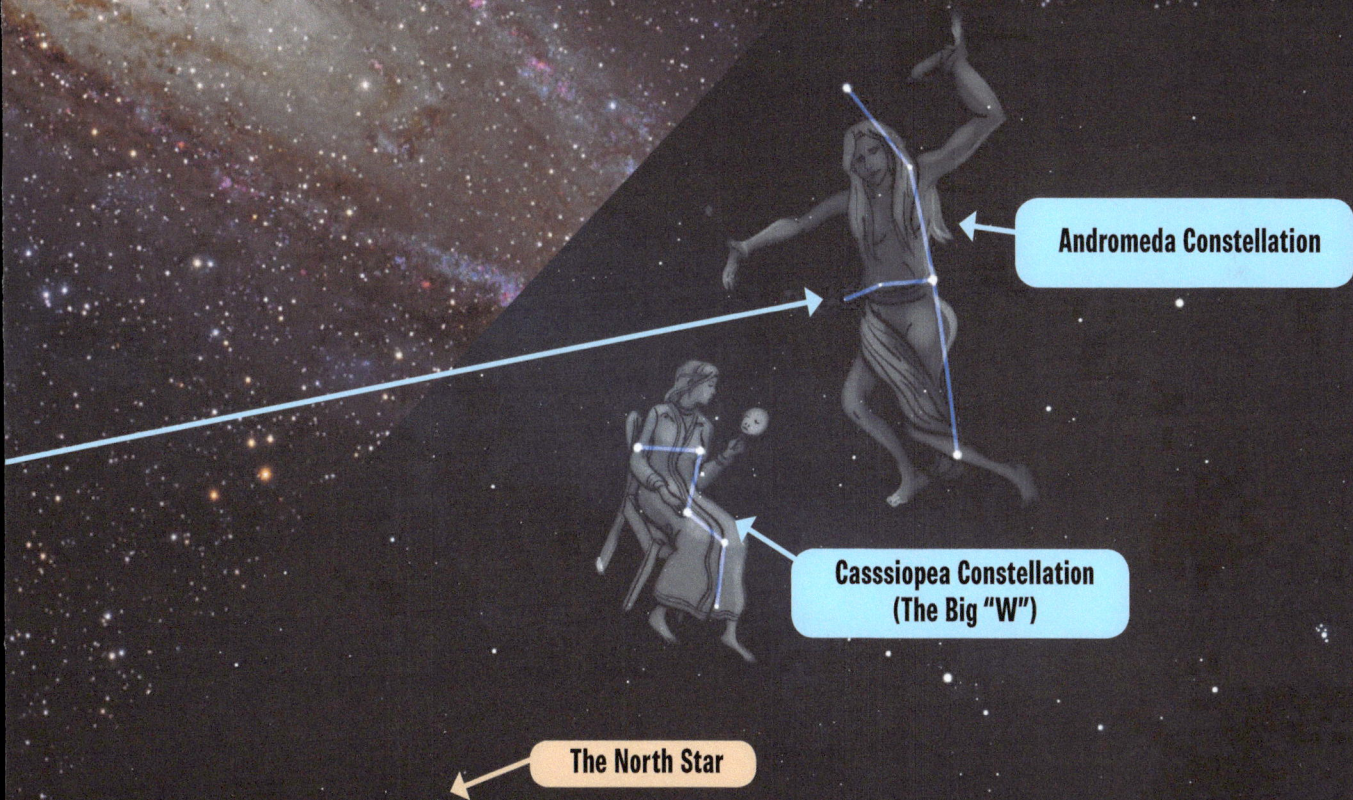

Andromeda Constellation

Casssiopea Constellation
(The Big "W")

The North Star

In the west, the Great Globular Cluster in Hercules begins its dive below the horizon.

Globular Clusters

Globular Clusters are tight group of thousands of stars orbiting our galaxy in a region called the Halo. If you are patient, and have dark skies, dozens of these objects can be seen with a small telescope.

TELESCOPE VIEW

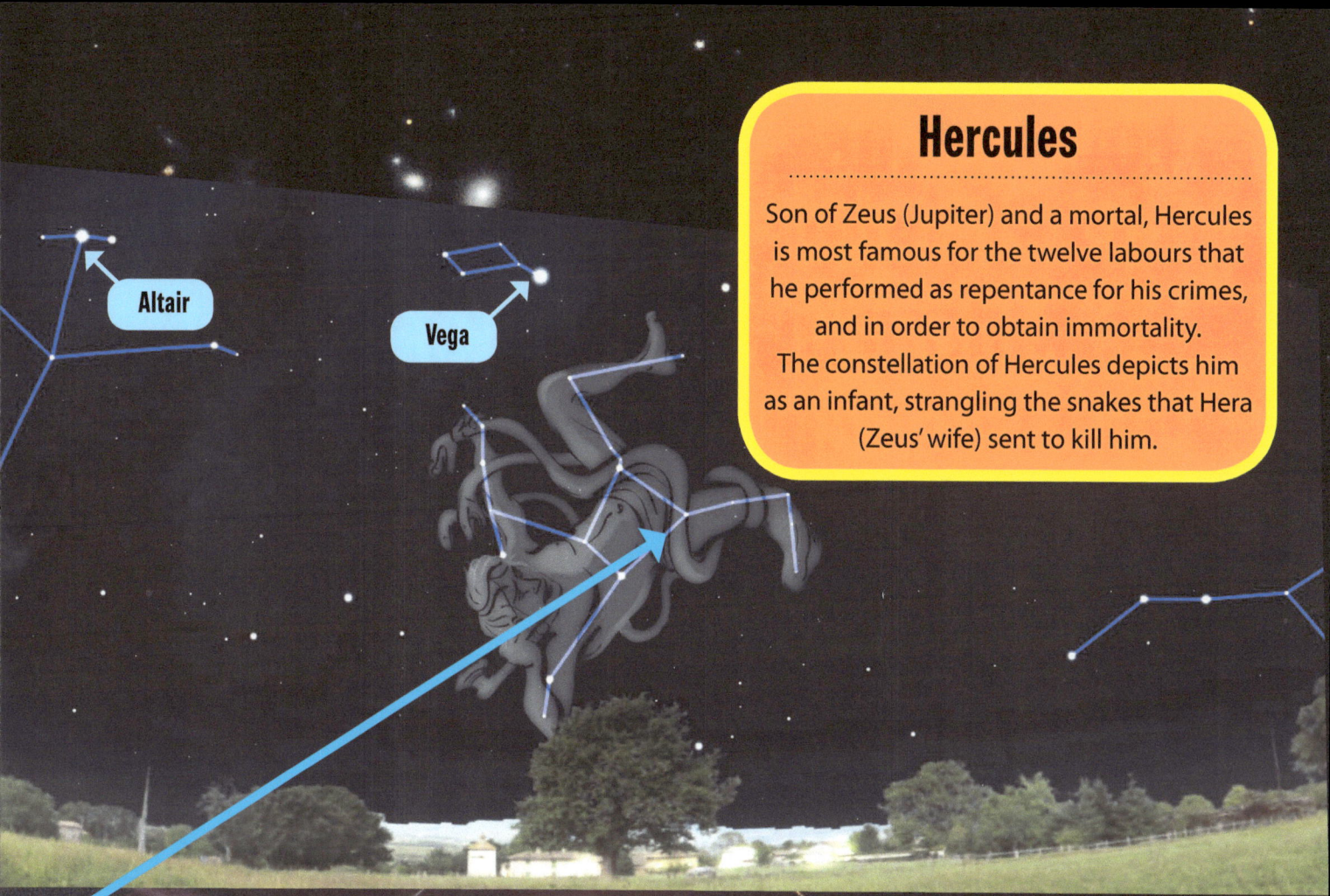

Altair

Vega

Hercules

Son of Zeus (Jupiter) and a mortal, Hercules is most famous for the twelve labours that he performed as repentance for his crimes, and in order to obtain immortality.
The constellation of Hercules depicts him as an infant, strangling the snakes that Hera (Zeus' wife) sent to kill him.

Supermassive Black Holes

These wisps of pink in the background image are radio emissions (light at radio wavelengths) detected by a radio telescope in New Mexico. They are called "Jets" and are made of subatomic particles thrust into the universe by the gravitation energy of a galaxy's central supermassive black hole.

Low on the Eastern horizon, the Pleiades slowly rises, heralding in the winter which will soon cover the north with snow.

Betelgeuse

Orion's Belt

Ceaseless Chase

According to Greek myth, the Pleiades, or seven sisters, were being pursued by the hunter Orion when they were placed in the sky in order to escape. Ironically, Orion continues to chase them, even there.

TELESCOPE VIEW

In Japan, The Pleiades go by the name "Subaru"

In the western sky, at the head of Cygnus the Swan, a golden star tugs on its neighbor, a blue giant, in a cosmic dance.

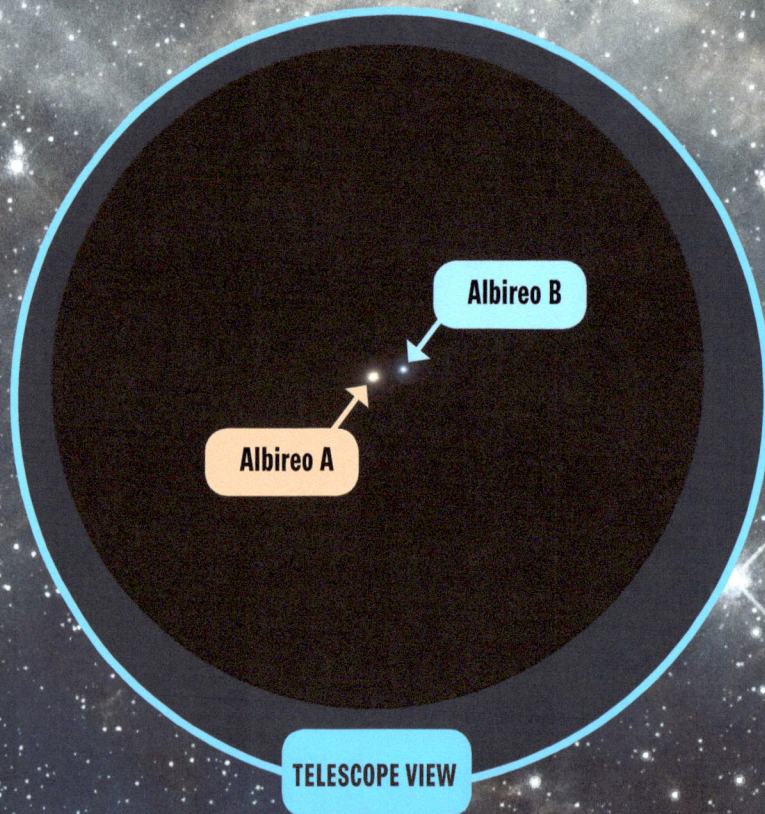

Albireo B

Albireo A

TELESCOPE VIEW

The "Northern Cross" in Cygnus

Vega

Cygnus the Swan

It is unclear who in myth Cygnus represents. One option is that it is Zeus (Jupiter) when he transformed into a swan to meet Leda. In myth, Zeus would often disguise himself as different animals when he visited humans.

High above us, a thousand tiny rocks penetrate the Earth's atmosphere at over 100,000 kilometers per hour. Iron, magnesium, and silicon vaporize, glowing with intense heat. Some of them explode, showering the night in fireballs that streak across the sky, leaving a trail of smoke in their wake.

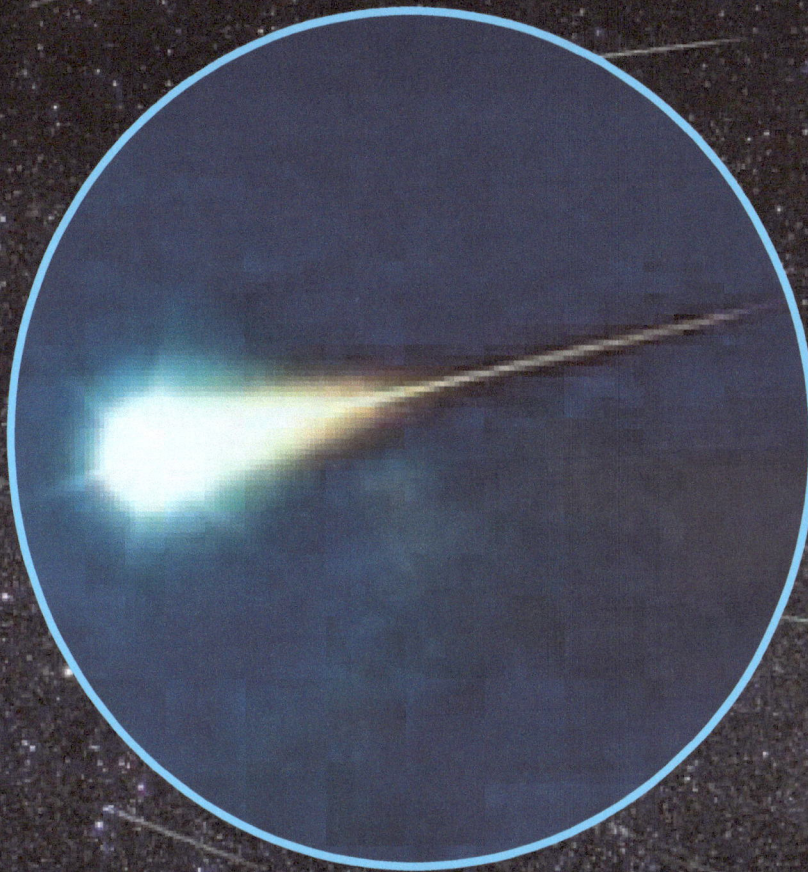

Meteoriod -> Meteor -> Meteorite

A small rock in space is a meteoriod. When a meteoroid enters Earth's atmosphere it becomes a meteor. If that meteor hits the ground, it becomes a Meteorite.

300 kilometers above the Earth, a top secret US Air Force Space Plane orbits at 24,000 kilometers per hour.

The unmanned x-37b vehicle was built by Boeing's secretive Phantom Works division. Despite the secrecy, this spacecraft has been photographed, through moderately sized backyard telescopes!